USEFUL CONSTANTS

Avogadro's number	6.02×10^{23} molecules · mol^{-1}
Gas constant (R)	$8.314\ J \cdot K^{-1} \cdot mol^{-1}$
Faraday ($\mathcal{F}$)	$96{,}485\ J \cdot V^{-1} \cdot mol^{-1}$
Kelvin (K)	°C + 273

KEY EQUATIONS

Henderson–Hasselbalch equation

$$\text{pH} = \text{p}K + \log \frac{[\text{A}^-]}{[\text{HA}]}$$

Michaelis–Menten equation

$$v_0 = \frac{V_{\max}[\text{S}]}{K_\text{M} + [\text{S}]}$$

Lineweaver–Burk equation

$$\frac{1}{v_0} = \left(\frac{K_\text{M}}{V_{\max}}\right)\frac{1}{[\text{S}]} + \frac{1}{V_{\max}}$$

Nernst equation

$$\mathcal{E} = \mathcal{E}^{\circ\prime} - \frac{RT}{n\mathcal{F}} \ln \frac{[\text{A}_{reduced}]}{[\text{A}_{oxidized}]} \quad \text{or} \quad \mathcal{E} = \mathcal{E}^{\circ\prime} - \frac{0.026\ \text{V}}{n} \ln \frac{[\text{A}_{reduced}]}{[\text{A}_{oxidized}]}$$

Thermodynamics equations

$$\Delta G = \Delta H - T\Delta S$$

$$\Delta G^{\circ\prime} = -RT \ln K_{eq}$$

$$\Delta G = \Delta G^{\circ\prime} + RT \ln \frac{[\text{C}][\text{D}]}{[\text{A}][\text{B}]}$$

$$\Delta G^{\circ\prime} = -n\mathcal{F}\Delta\mathcal{E}^{\circ\prime}$$

Common Functional Groups and Linkages in Biochemistry

Compound Name	Structure[a]	Functional Group
Amine[b]	RNH_2 or RNH_3^+ R_2NH or $R_2NH_2^+$ R_3N or R_3NH^+	—N< or —$\overset{+}{N}$— (amino group)
Alcohol	ROH	—OH (hydroxyl group)
Thiol	RSH	—SH (sulfhydryl group)
Ether	ROR	—O— (ether linkage)
Aldehyde	R—C(=O)—H	—C(=O)— (carbonyl group), R—C(=O)— (acyl group)
Ketone	R—C(=O)—R	—C(=O)— (carbonyl group), R—C(=O)— (acyl group)
Carboxylic acid[b] (Carboxylate)	R—C(=O)—OH or R—C(=O)—O^-	—C(=O)—OH (carboxyl group) or —C(=O)—O^- (carboxylate group)
Ester	R—C(=O)—OR	—C(=O)—O— (ester linkage)
Amide	R—C(=O)—NH_2 R—C(=O)—NHR R—C(=O)—NR_2	—C(=O)—N< (amido group)
Imine[b]	R=NH or R=NH_2^+ R=NR or R=NHR^+	>C=N— or >C=$\overset{+}{N}$<H (imino group)
Phosphoric acid ester[b]	R—O—P(=O)(OH)—OH or R—O—P(=O)(O^-)—O^-	—O—P(=O)(OH)—O— (phosphoester linkage) —P(=O)(OH)—OH or —P(=O)(O^-)—O^- (phosphoryl group, P_i)
Diphosphoric acid ester[b]	R—O—P(=O)(OH)—O—P(=O)(OH)—OH or R—O—P(=O)(O^-)—O—P(=O)(O^-)—O^-	—O—P(=O)(OH)—O—P(=O)(OH)—O— (phosphoanhydride linkage) —P(=O)(OH)—O—P(=O)(OH)—OH or —P(=O)(O^-)—O—P(=O)(O^-)—O^- (diphosphoryl group, pyrophosphoryl group, PP_i)

[a] R represents any carbon-containing group. In a molecule with more than one R group, the groups may be the same or different.
[b] Under physiological conditions, these groups are ionized and hence bear a positive or negative charge.

Essential Biochemistry

Fourth Edition

CHARLOTTE W. PRATT
Seattle Pacific University

KATHLEEN CORNELY
Providence College

WILEY

VICE PRESIDENT, PUBLISHER	Petra Recter
SPONSORING EDITOR	*Joan Kalkut*
ASSOCIATE DEVELOPMENT EDITOR	Laura Rama
SENIOR MARKETING MANAGER	Kristine Ruff
PRODUCT DESIGNER	Sean Hickey
SENIOR DESIGNER	Thomas Nery
SENIOR PHOTO EDITOR	Billy Ray
EDITORIAL ASSISTANT	Mili Ali
PRODUCT DESIGN ASSISTANT	Alyce Pellegrino
MARKETING ASSISTANT	Maggie Joest
SENIOR PRODUCTION EDITOR	Elizabeth Swain

Cover image: The active site of RNA polymerase, based on a structure determined by Yuan He, Chunli Yan, Jie Fang, Carla Inouye, Robert Tjian, Ivaylo Ivanov, and Eva Nogales (pdb 5IYD).

This book was typeset by Aptara and printed and bound by Quad Graphics Versailles. The cover was printed by Quad Graphics Versailles.

ISBN 978-1-119-31933-7

The inside back cover will contain printing identification and country of origin if omitted from this page. In addition, if the ISBN on the back cover differs from the ISBN on this page, the one on the back cover is correct.

Printed in the United States of America

V007101_052118